Project M^3: Mentoring Mathematical Minds Series

At the Mall With Algebra

Working With Variables and Equations

M. Katherine Gavin, *University of Connecticut*
Suzanne H. Chapin, *Boston University*
Judith Dailey, *Connecticut Association for Mathematically Precocious Youth*
Linda Jensen Sheffield, *Northern Kentucky University*

Student Mathematician's Journal

Common Core State Standards

Kendall Hunt
publishing company

www.kendallhunt.com
Send all inquiries to:
4050 Westmark Drive
Dubuque, IA 52002
1-800-542-6657

The contents of this book were developed under a grant from the Department of Education, Jacob K. Javits Gifted and Talented Students Education Act. However, those contents do not necessarily represent the policy of the Department of Education, and you should not assume endorsement by the Federal Government.

ISBN 978-1-5249-2852-0

Printed in the United States of America

1 2 3 4 5 6 7 8 9 10

Production Date: 2016
Printed by LSI

United States of America
Batch number: 436214

TABLE OF CONTENTS

CLASSROOM DISCUSSIONS

You have the right to ask questions.

You have the right to make a contribution to an attentive, responsive audience.

You have the right to be treated respectfully.

You have the right to have your ideas discussed, not you.

Rights

You are obligated to speak loudly enough for others to hear.

You are obligated to listen to others in order to understand.

You are obligated to agree or disagree with the speaker's comments and explain why.

Obligations

Adapted from Chapin, S.H., O'Connor, C., & Anderson, N. C. (2013). *Classroom discussions using math talk to help students learn, grades K–6* (3rd ed.). Sausalito, CA: Math Solutions Publications.

THINKING LIKE A MATHEMATICIAN*

Here is a list of skills mathematicians use every day. See how many you can use in your Student Mathematician's Journal.

1. Make sense of problems and keep trying until you solve them.
2. Understand quantities, their relationships, and how to represent them.
3. Build logical reasons to defend your thinking. Consider the reasoning of others and ask useful questions to help make sense of the reasoning. Explain why you agree or disagree with another's reasoning.
4. Use the math you know to help solve problems in everyday life. Use physical models, drawings, tables, graphs, and/or equations to help you.
5. Choose and use the appropriate math tools to help solve each problem.
6. Communicate explanations clearly using correct math vocabulary and symbols.
7. Look closely and use patterns to help solve problems.
8. Notice if you are using the same math again and again and look for short cuts.
9. Solve a problem in a new way. Ask new questions to investigate.**

* Adapted from the Common Core State Standards: Standards for Mathematical Practice

National Governors Association Center for Best Practices (NGA Center), Council of Chief State School Officers (CCSSO). (2010). *Common Core State Standards for Mathematics.* Washington, DC: Retrieved from http://www.corestandards.org/the-standards.

** Johnsen, S. K., & Sheffield, L. J. (Eds.). (2013). *Using the common core state standards for mathematics with gifted and advanced learners.* Waco, TX : Prufrock Press.

MATHEMATICIANS WRITE ON!

When you write in your Mathematician's Journal:

1. Read the questions. Make sure you understand every part and vocabulary word in the questions.
2. Brainstorm ideas.
3. Write your answer.
4. Read over your answers and check that you:

   a. answered all of the questions;
   b. defended your answers by using vocabulary words, drawings, numbers, and/or other details.

5. Revise your answers if you think someone else reading it would not know what you were thinking by asking yourself the following questions:

   a. "Is this clear?"
   b. "Will the reader understand why I did what I did?"

Student Mathematician: ______________________ Date: __________

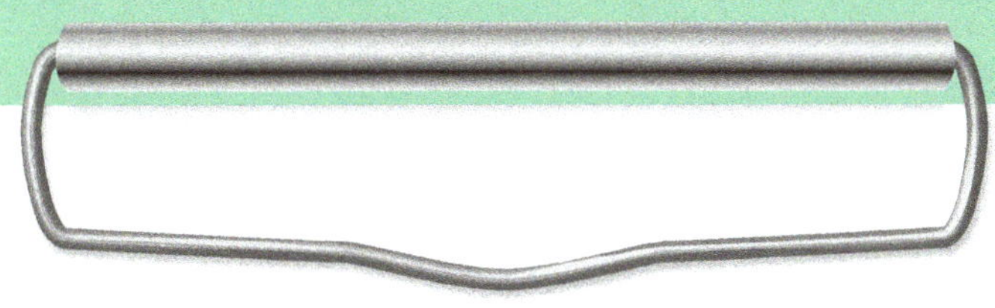

1. Think of a number. Apply this rule to your number: 2 times the number, plus 3.

   a. What is your answer?

   b. Explain how someone else could figure out your original number by working backwards to undo operations.

MY THOUGHTS AND QUESTIONS

MY RESPONSE

Need more room? Use the next page.

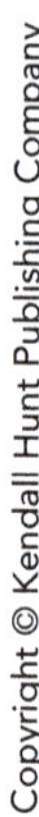

Student Mathematician: ______________________ Date: ____________

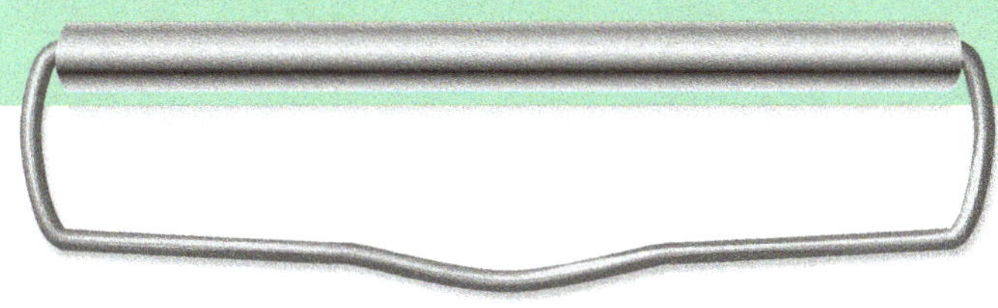

2. a. In your own words, define the terms *expression* and *equation*.

b. Give an example of an expression and an equation.

c. How are they alike?

d. How are they different?

MY THOUGHTS AND QUESTIONS

MY RESPONSE

Need more room? Use the next page.

Student Mathematician: ______________________ Date: __________

Mathematical Trickery!

Part I: Make the following cards for the math trick.

	Card A	Card B	Card C	Card D	Card E
Front	1	3	5	7	9
Back	2	4	6	8	10

1. If you used only Card A and Card B to play the game, what is the minimum sum possible? __________ Which numbers give the minimum sum? __________ What is the maximum possible sum? __________ Which numbers give the maximum sum? __________

2. If you used only Card A, Card B and Card C to play the game, what is the minimum sum possible? __________Which numbers give the minimum sum? __________ What is the maximum sum possible? __________ Which numbers give the maximum sum? __________

3. What is special about the numbers used to get the minimum and maximum sums?

____

____

4. What number patterns do you see on a) the fronts? b) the backs? c) the front to back?

____

____

5. The person performing the number trick always asks to be told how many even numbers are used. Why do you think you have to know how many even numbers are used in order to predict the sum?

____

____

Mathematical Trickery! (continued)

6. Explain how the number trick works.

7. Write a rule using variables for how to find the sum of the numbers on the cards.

Part II: Make your own set of number trick cards.

8. Make up a set of cards to use for the number trick. You may use any number of cards. Indicate here what numbers are on the fronts and backs of your cards.

   Front:

   Back:

9. Write the directions for your number trick.

10. What is the minimum sum possible using your set of cards? ____________

11. What is the maximum sum possible using your set of cards? ____________

12. What question will you ask that will give you more information about the sum?

13. Write a rule with variables that can be used to determine the sum of your numbers.

Student Mathematician: ______________________ Date: ____________

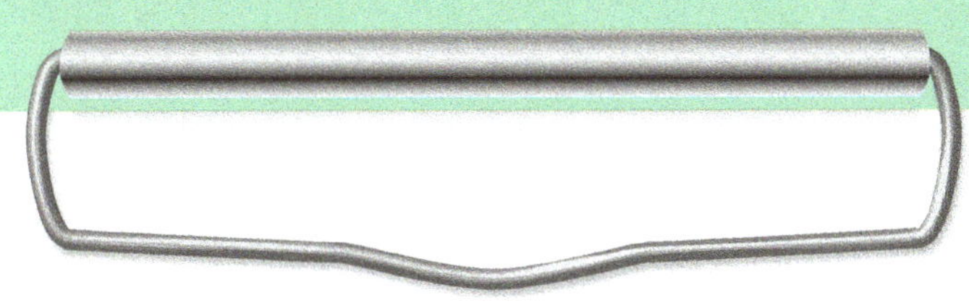

1. Your best friend missed school but heard that you learned a number trick in math class. You decide to send your friend your set of cards and a note explaining how your number trick works.

   a. What numbers were on your cards?

   b. What was your rule?

   c. Why does your rule work?

MY RESPONSE

Need more room? Use the next page.

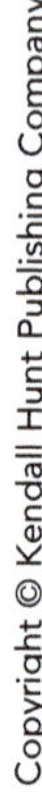

Student Mathematician: ______________________ Date: __________

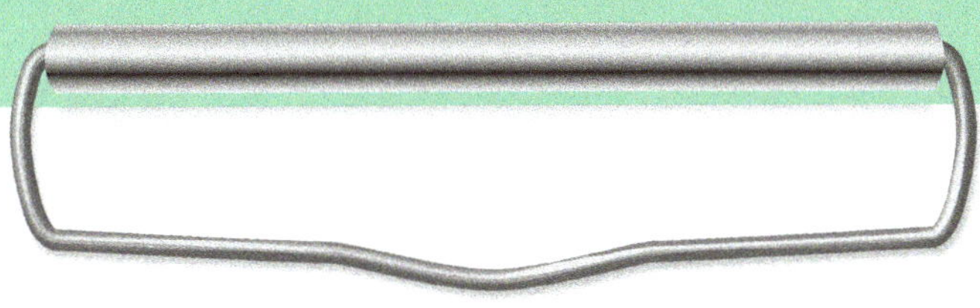

2. a. What is a variable?

   b. How did you use a variable to write a rule for your card trick?

MY THOUGHTS AND QUESTIONS

MY RESPONSE

Need more room? Use the next page.

Student Mathematician: ______________________ Date: ____________

Variable Puzzles

The same variables (letters) have the same value everywhere in an individual puzzle, but the value may change from one puzzle to another. Different variables in a puzzle have different values. The numbers at the end of each row and column are the sums of all the values in that row or column. Find the value for each variable.

1	Column 1	Column 2	Column 3	Row Total
Row 1	*D*	9	*C*	17
Row 2	*D*	*D*	10	22
Row 3	3	*C*	*C*	7
Column Total	15	17	14	

C = ______

D = ______

2	Column 1	Column 2	Column 3	Row Total
Row 1	7	*F*	8	24
Row 2	*E*	*F*	3	16
Row 3	*F*	*F*	*E*	22
Column Total	20	27	15	

E = ______

F = ______

3	Column 1	Column 2	Column 3	Row Total
Row 1	4	*G*	*H*	27
Row 2	*G*	*G*	*H*	34
Row 3	*G*	7	*H*	30
Column Total	26	29	36	

G = ______

H = ______

4	Column 1	Column 2	Column 3	Row Total
Row 1	6	*J*	*J*	12
Row 2	8	*K*	*K*	16
Row 3	*N*	*N*	*K*	18
Column Total	21	14	11	

J = ______

K = ______

N = ______

Student Mathematician: ______________________ Date: ____________

Variable Puzzles (continued)

5	Column 1	Column 2	Column 3	Row Total
Row 1	*L*	3	*M*	22
Row 2	5	*M*	9	27
Row 3	*P*	*P*	7	27
Column Total	21	26	29	

L = ______
M = ______
P = ______

6	Column 1	Column 2	Column 3	Row Total
Row 1	*R*	*Q*	*S*	12
Row 2	*S*	*R*	9	20
Row 3	*S*	*S*	8	26
Column Total	20	12	26	

Q = ______
R = ______
S = ______

7	Column 1	Column 2	Column 3	Row Total
Row 1	*Z*	8	*W*	11
Row 2	7	*Z*	*W*	10
Row 3	*X*	*Y*	*Z*	9
Column Total	13	13	4	

W = ______
X = ______
Y = ______
Z = ______

8	Column 1	Column 2	Column 3	Row Total
Row 1	*A*	4	*B*	16
Row 2	*B*	13	2	25
Row 3	7	*C*	*A*	15
Column Total	19	23	14	

A = ______
B = ______
C = ______

Student Mathematician: ______________________________ Date: ____________

Create Your Own Variable Puzzles

The same variables (letters) have the same value everywhere in an individual puzzle, but the value may change from one puzzle to another. Different variables in a puzzle have different values. The numbers at the end of each row and column are the sums of all the values in that row or column. Find the value for each variable.

	Column 1	Column 2	Column 3	Row Total
Row 1				
Row 2				
Row 3				
Column Total				

	Column 1	Column 2	Column 3	Row Total
Row 1				
Row 2				
Row 3				
Column Total				

	Column 1	Column 2	Column 3	Row Total
Row 1				
Row 2				
Row 3				
Column Total				

	Column 1	Column 2	Column 3	Row Total
Row 1				
Row 2				
Row 3				
Column Total				

Student Mathematician: ______________________ Date: __________

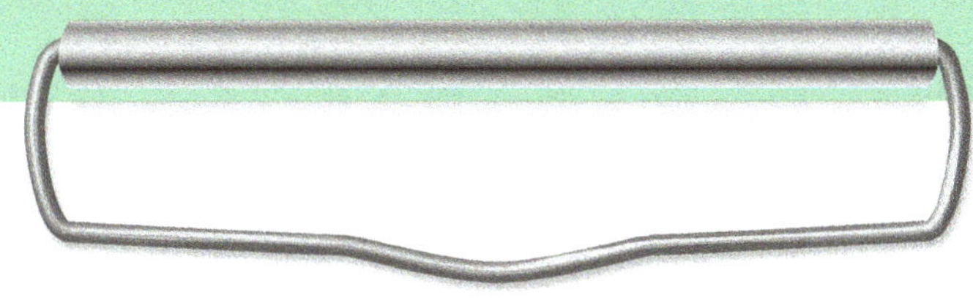

1. Copy one of the puzzles from "Variable Puzzles" into your Mathematician's Journal. Describe how you figured out the values of the variables.

MY THOUGHTS AND QUESTIONS

MY RESPONSE

Need more room? Use the next page.

Student Mathematician: ______________________ Date: ____________

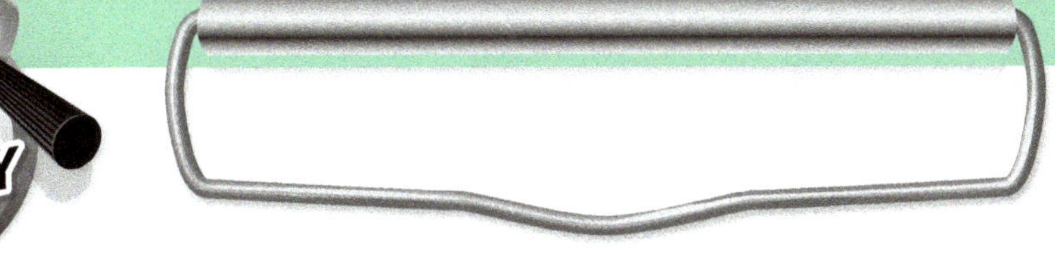

2. Julia mentioned that she found the value of the letter X in row 1 in the puzzle below using a guess-and-test strategy, whereas Peter said he found the value by subtracting and dividing. Do both strategies work? Explain.

	Column 1	Column 2	Column 3	Row Total
Row 1	X	X	25	41

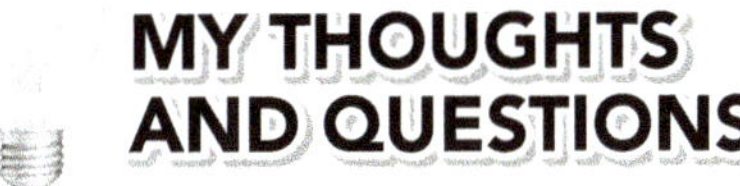

MY RESPONSE

Need more room? Use the next page.

Cover Up!

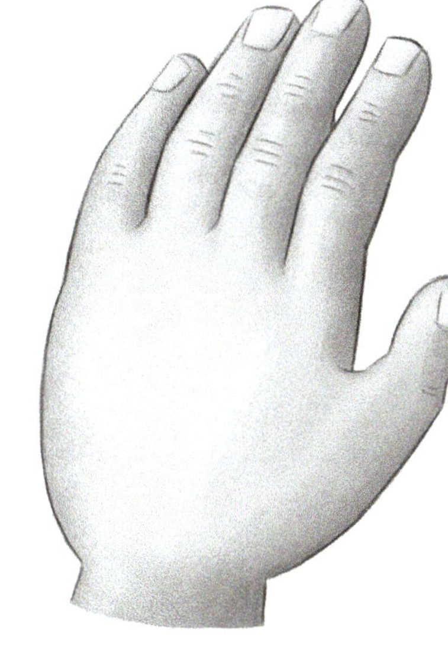

Solve these equations using the cover-up method. Lightly shade what you covered up and box the value of the covered expression, as in the example below.

Cover up	$2N + 7 = 15$	
	$\boxed{8} + 7 = 15$	*Let's check!*
Uncover	$2N = 8$	$2N + 7 = 15$
Cover up	$2 \bullet N = 8$	$2 \bullet 4 + 7$
	$2 \bullet \boxed{4} = 8$	$8 + 7$
Uncover	$N = 4$	15

Yes! We're correct!

1. $N + 16 = 40$

2. $12 = 3N \div 2$

3. $4N - 6 = 94$

4. $12 + 2N = 34$

Student Mathematician: ______________________ Date: __________

Cover Up! (continued)

5. $6 = 9N \div 3$

6. $6N - 156 = 600$

7. $7N + \frac{1}{2} = 14\frac{1}{2}$

8. $(N \div 2) + 7 = 11$

9. $(N \div 5) - 8 = 22$

10. $120 + 2N = 340$

11. $3N \div 8 = 3$

12. $4{,}000 = 216 + N$

Student Mathematician: ______________________ Date: __________

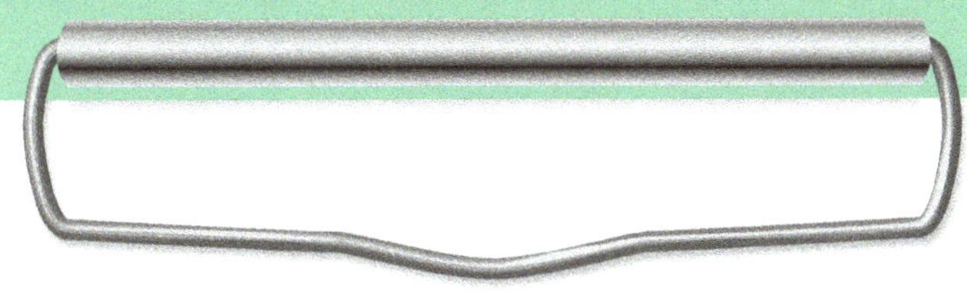

1. Choose an equation from the "Cover Up!" worksheet and describe how to solve it using the cover-up method.

MY RESPONSE

Need more room? Use the next page.

Student Mathematician: ______________________ Date: __________

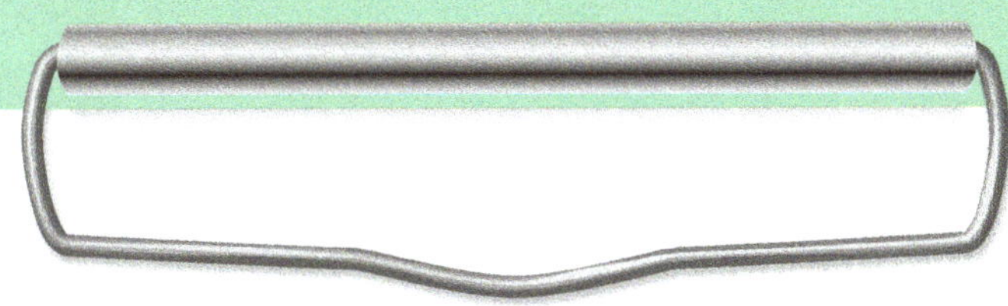

2. Here is an equation: $7 + 2x = 9$.

a. Explain two different ways to determine the value of x.

b. What is the value of x?

c. Which method do you think is better? Why?

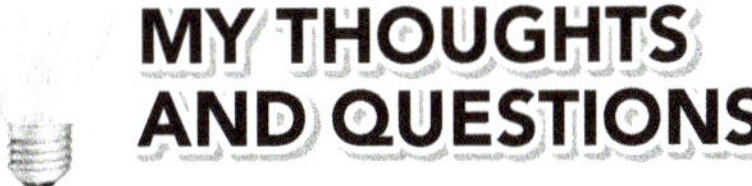

MY RESPONSE

Need more room? Use the next page.

Student Mathematician: ______________________ Date: ____________

Pet Parade

Find the number of each kind of pet for the following problems. Use the guess-and-test method with an organized list to help you solve problems 1–3. Draw a picture to help you solve the rest of the problems.

1. Hannah and Timothy have some dogs and African birds as their pets. There are 8 animals in all. Hannah counted their legs and found that there were a total of 24 legs. How many dogs and how many African birds did they have? *(Problem contributed by Rachel, a fourth grade student)*

Student Mathematician's Workspace

What We Know

What We Want to Find Out

Organized List

Guess 1:

Test 1:

Guess 2:

Test 2:

Answer:

Student Mathematician: ______________________ Date: __________

Pet Parade (continued)

2. Harry and his friends have an interesting collection of animals. The animals have a total of 34 legs. The animals, as you might have guessed, are owls and rats, and their total number is 11. How many of each do they have?

Student Mathematician's Workspace

What We Know

What We Want to Find Out

Organized List

Guess 1:

Test 1:

Guess 2:

Test 2:

Answer:

Student Mathematician: ______________________ Date: ____________

Pet Parade (continued)

3. Ken and Kerry live on a farm with lots of animals. Their friend Kiara came to visit and counted the total number of pigs and ducks. There were 19 animals with a total of 54 legs. How many pigs and how many ducks were there?

Student Mathematician's Workspace

What We Know

What We Want to Find Out

Organized List

Guess 1:

Test 1:

Guess 2:

Test 2:

Answer:

Student Mathematician: ______________________ Date: __________

Pet Parade (continued)

4. Serena loves spiders and fish. She has 10 animals. She counted the legs on all the animals and came up with 24 legs. How many of each animal did she have?
(This one is tricky!)

Student Mathematician's Workspace

What We Know	What We Want to Find Out
Picture	
Answer:	

Student Mathematician: ______________________ Date: ____________

Pet Parade (continued)

5. Wanda plans to become a veterinarian and loves to collect ants and frogs. She has 5 animals with a total of 24 legs. How many of each animal does she have?

Student Mathematician's Workspace

What We Know	**What We Want to Find Out**

Picture

Answer:

Pet Parade (continued)

6. In the Land of Treble, everything comes in groups of three, so every animal has three heads! One day, the mayor of the Land of Treble counted 27 heads on the animals as she walked through the pet store. This pet store sells three kinds of animals: gurkles, burkles and turkles. There happened to be 9 heads in the store that day that belonged to turkles, and there were also twice as many gurkles as burkles. How many gurkles, burkles and turkles were in the store?

Student Mathematician's Workspace

What We Know	**What We Want to Find Out**
Picture	
Answer:	

Student Mathematician: ______________________ Date: __________

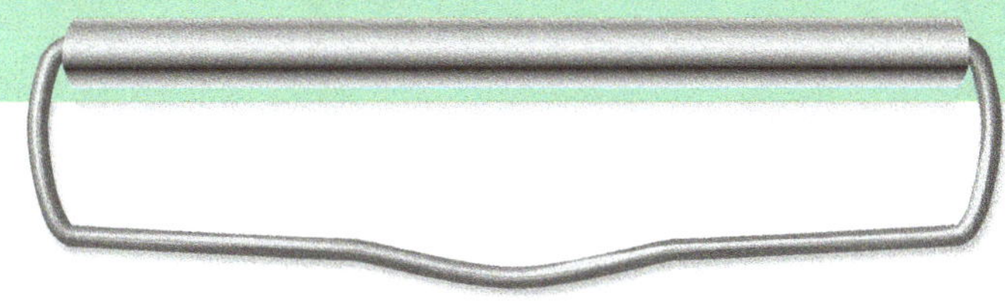

1. Explain how you solved the following problem using the guess-and-test method. Make sure you explain how using an organized list helped you make good guesses.

Ken and Kerry live on a farm with lots of animals. Their friend Kiara came to visit and counted the total number of pigs and ducks. There were a total of 19 animals with a total of 54 legs. How many pigs and how many ducks were there?

MY RESPONSE

Need more room? Use the next page.

Student Mathematician: ______________________ Date: __________

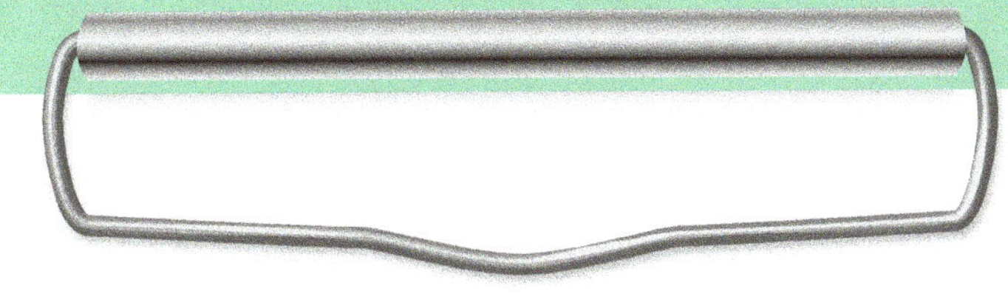

2. Explain how you solved the following problem using a picture. Make sure you explain how you used each of the numbers in the puzzle to help you.

Wanda plans to become a veterinarian and loves to collect ants and frogs. She has a total of 5 animals with a total of 24 legs. How many of each does she have?

MY THOUGHTS AND QUESTIONS

MY RESPONSE

Need more room? Use the next page.

Student Mathematician: ______________________________ Date: ____________

A Penny for Your Thoughts

Here are some signs found at the penny candy counter at the Mayberry General Store. Can you figure out what the value of each candy is? Make sure you check your answers.

1.

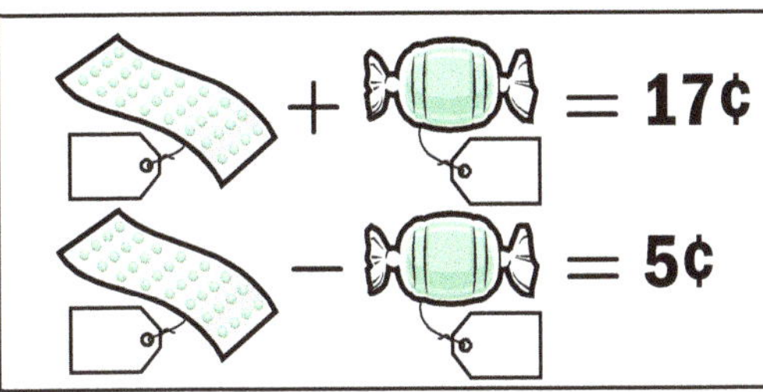

Value of dot candy strip = ________

Value of root beer barrel = ________

Student Mathematician's Workspace

Check:

2.

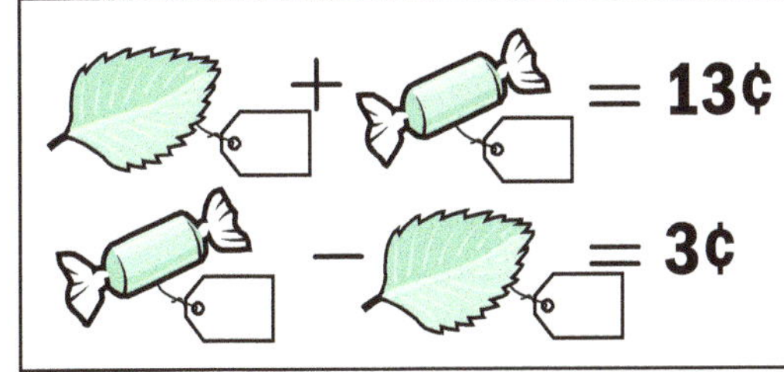

Value of Tootsie Roll® = ________

Value of spearmint leaf = ________

Student Mathematician's Workspace

Check:

3.

Value of dot chocolate bar = ________

Value of orange slice = ________

Student Mathematician's Workspace

Check:

A Penny for Your Thoughts (continued)

4.

+ = 32¢

− = 18¢

Value of peppermint stick = ________

Value of gummy bear = ________

Student Mathematician's Workspace

Check:

5.

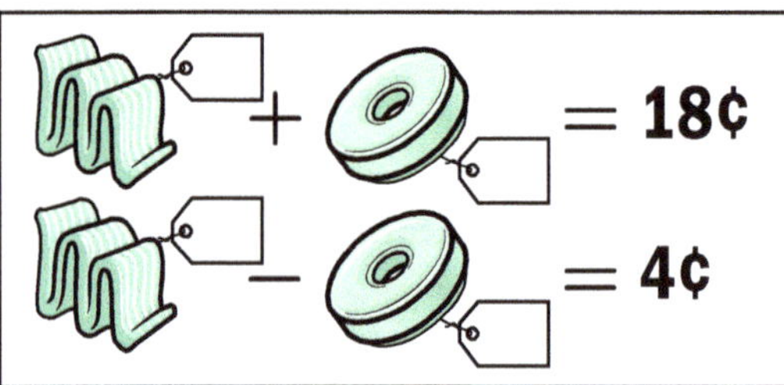

Value of piece of taffy = ________

Value of Lifesavers® = ________

Student Mathematician's Workspace

Check:

6.

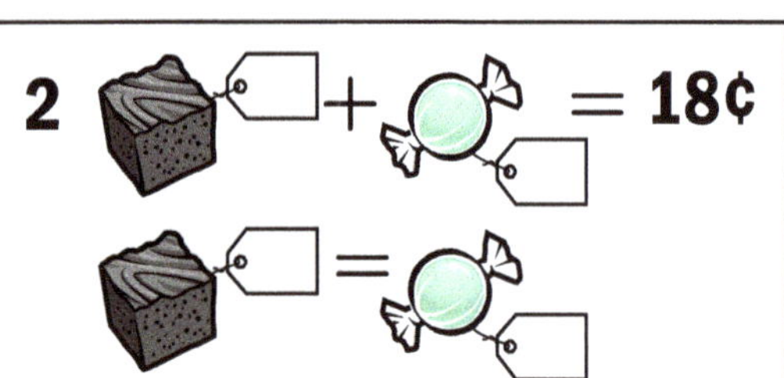

Value of piece of fudge = ________

Value of fireball = ________

Student Mathematician's Workspace

Check:

Student Mathematician: ______________________ Date: __________

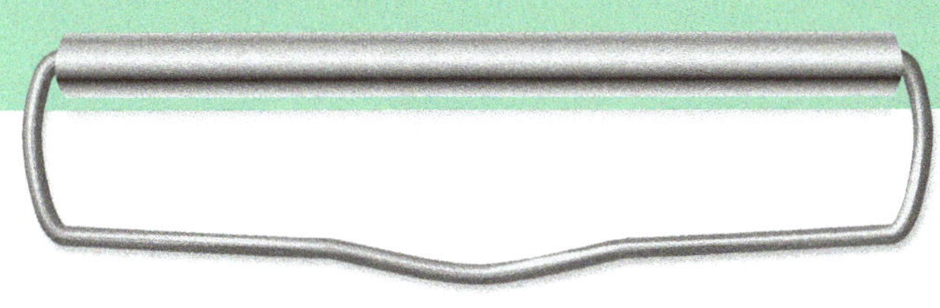

1. A new sign just went up at the Mayberry General Store. What do the candies cost?

As a first guess, James thought he would use 16¢ for the caramel and 12¢ for the squirrel nut. Jane thought she would try 20¢ for the caramel and 8¢ for the squirrel nut.

a. Which is the better first guess?

b. How can you tell without checking if the numbers work?

MY THOUGHTS AND QUESTIONS

Need more room? Use the next page.

MY RESPONSE

Student Mathematician: ______________________ Date: ____________

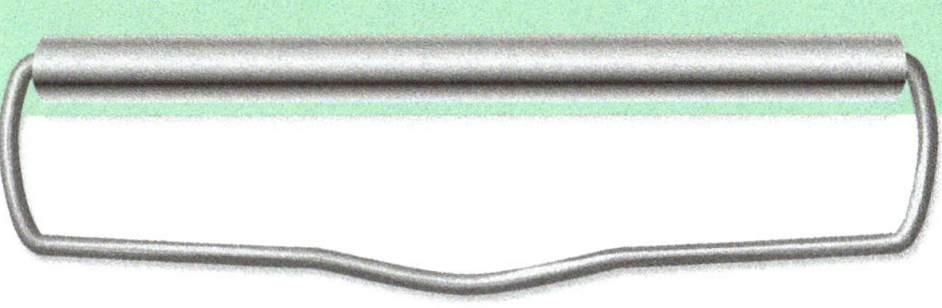

2.

[dot candy strip] + [root beer barrel] = **17¢**

[dot candy strip] − [root beer barrel] = **1¢**

a. What is the value of the dot candy strip and of the root beer barrel?

b. Explain how you got your answers.

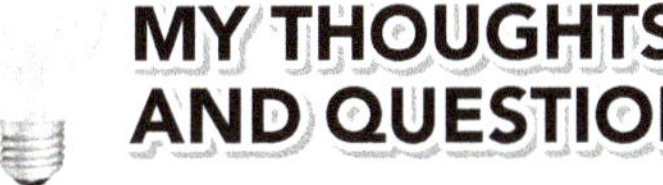

MY THOUGHTS AND QUESTIONS

MY RESPONSE

Need more room? Use the next page.

Student Mathematician: ______________________ Date: __________

What's the Weather?

Find the number that goes inside each symbol to make the equations true and write it on the blank line. Same symbols have the same number value in each problem.

1. ___ + 7 = 20

___ + ___ = 28

2. 16 − ___ = ___

___ + 7 − 2 = 11

3. ___ − ___ = 21

___ ÷ 6 = 5

4. ___ + ___ = 24

___ × 3 = 24

5. ___ = ___ + 2

___ − 7 = 2

Student Mathematician: ______________________ Date: ____________

Seasonal Symbols

Find the number that goes inside each symbol to make the equations true and write it in the white square. Same symbols have the same number value in each problem. Be sure to check your answers in both equations.

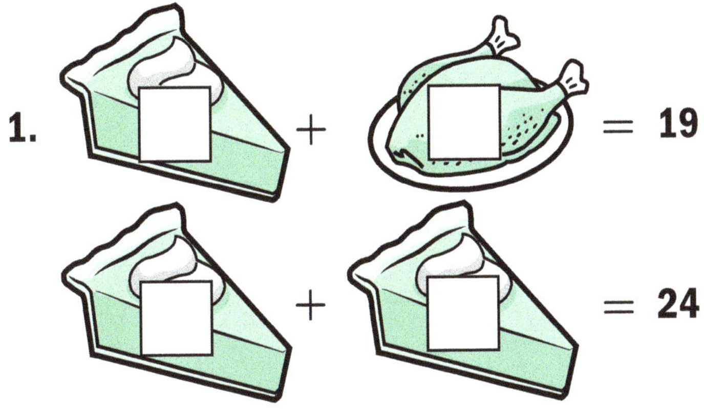

Which value did you find first in problem 2 — heart or flower? Explain why you started with this variable.

____

____

____

Student Mathematician: ______________________ Date: __________

Seasonal Symbols (continued)

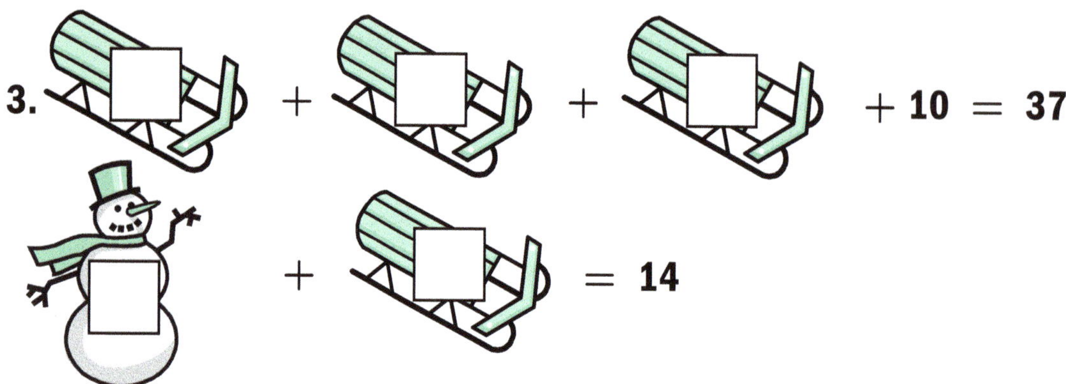

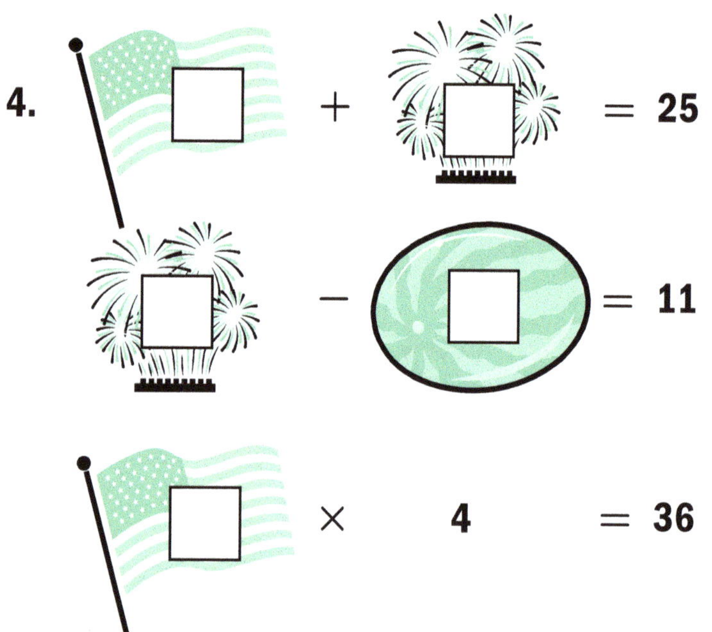

Show how you checked your answers for problem 4.

____

____

____

Seasonal Symbols (continued)

Which value did you find first in problem 7? Explain how you found it.

____

____

____

Seasonal Symbols (continued)

8.

9.

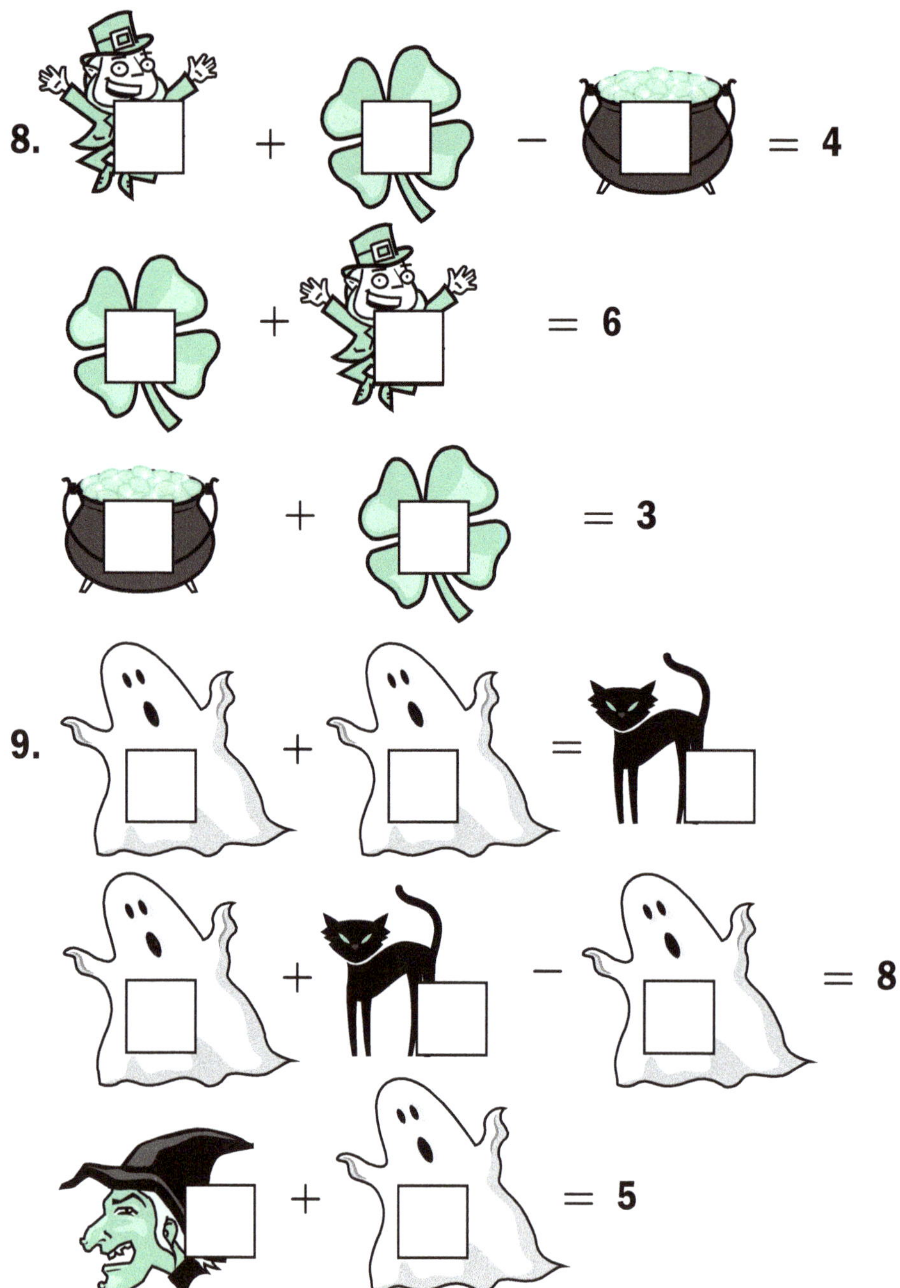

For problem 9, why is it a good idea to start with the second equation?

Student Mathematician: ______________________ Date: __________

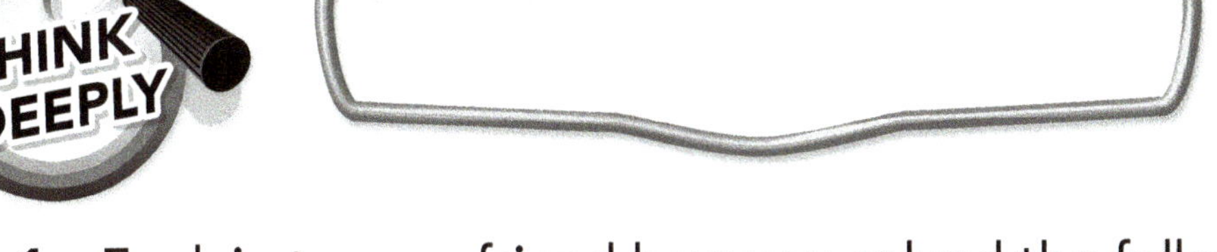

1. Explain to your friend how you solved the following problem from "Seasonal Symbols":

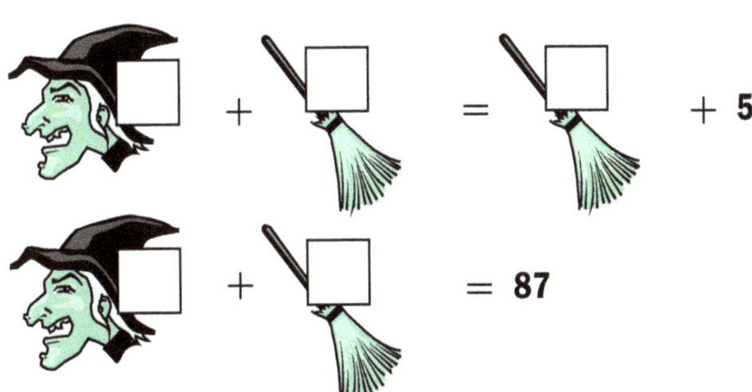

MY RESPONSE

Need more room? Use the next page.

Student Mathematician: ____________________ Date: __________

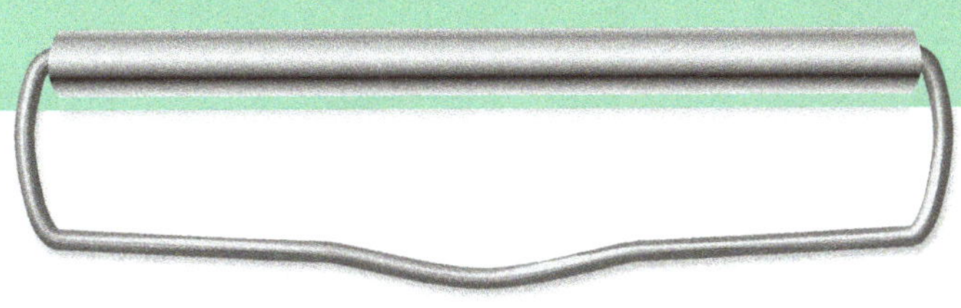

2. Kendra said the value of Lincoln was 6 in the following problem:

Explain to her why her answer is incorrect.

MY RESPONSE

Need more room? Use the next page.

Student Mathematician: ______________________ Date: ____________

Setting Up Shop

Using the three items you are advertising for your store, write the following:

1. A single equation with one item as a variable.

   For example: 4  − 3 = $17

2. A set of two equations, each with two items as variables.

   For example:  +  = $20

   −  = $4

Remember to make your pictures of the objects look as realistic as possible. For example, you can use clip art from a computer, stickers, magazine cut-outs, or good drawings.

Be sure to include the solutions on the back.

Store: ______________________

Store Owner's Name(s): ______________________

____

1. single equation advertising one item:

2. set of two equations advertising two other items:

Student Mathematician: ____________________ Date: ____________

Shopping at the Mall

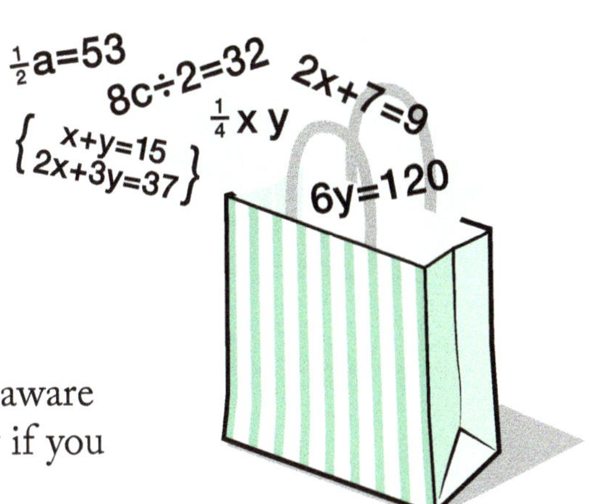

- Visit five stores before returning to work at your own store.
- You must solve one single-variable equation and one set of equations at each store that you visit.
- *Fill out the receipt below as you visit each store!* Be aware that the store owner will initial your receipt only if you solved the equations correctly!

Name of Store	Items Purchased	Cost of each Item	Store Owner's Initials
	1. 2. 3.	1. 2. 3.	
	1. 2. 3.	1. 2. 3.	
	1. 2. 3.	1. 2. 3.	
	1. 2. 3.	1. 2. 3.	
	1. 2. 3.	1. 2. 3.	

Student Mathematician: ______________________ Date: ____________

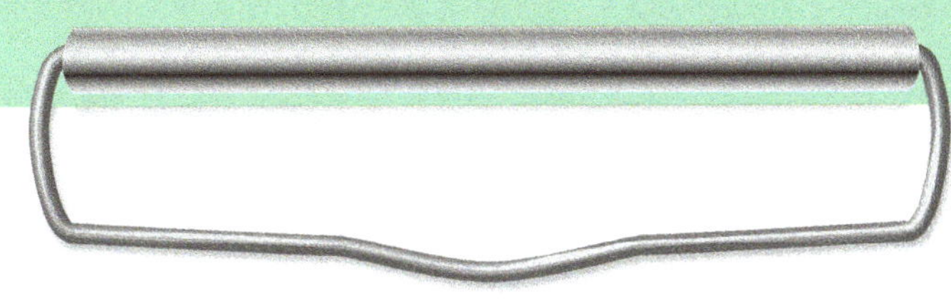

1. The word *equation* comes from "equate" which means "to make two things equal." What does this have to do with solving equations? Use an equation to help you explain your answer.

MY THOUGHTS AND QUESTIONS

MY RESPONSE

Need more room? Use the next page.

Student Mathematician: ______________________ Date: __________

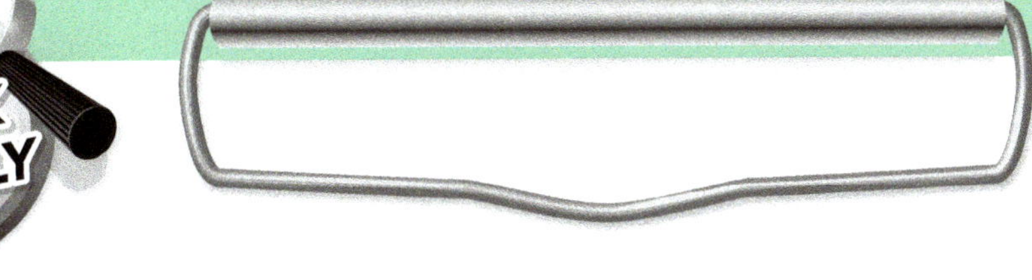

2. Pretend that a new student has just entered your class and wants to learn about your experience "At the Mall." Explain to her what a variable is and how we use them to solve equations. Use one of your mall ads to help you explain.

MY THOUGHTS AND QUESTIONS

MY RESPONSE

Need more room? Use the next page.

GLOSSARY

Equation a statement that says the value of an expression is equal to the value of another expression. For example, $7 = 7$, $8 = x$ and $20 = 2a + 6$ are equations.

Expression a statement involving numbers, operations and/or variables. Expressions do not use an equal sign. For example, $3 - 1$, x, $N \div 4 + 2$ and 6 are expressions.

Factor a number that divides exactly into another. For example, 7 is a factor of 14.

Guess-and-test method a problem-solving method in which one makes a guess, tests it to see if it answers the problem, and then refines the guess based on the test

Inverse operations operations that undo each other. For instance, addition and subtraction are inverse operations. For example, $5 + 4 = 9$, therefore, $9 - 4 = 5$. Multiplication and division also are inverse operations.

Organized list a listing of possible combinations written in an organized manner, for example, a list of numbers from least to greatest

Prime a number with two unique factors, 1 and itself. For example, 7 is a prime number since it has exactly two factors, 1 and 7. One is not prime since it has only one factor (1), not two unique factors.

Substitution method a method of solving a set of equations by finding the value of one variable using one equation and then using this value in place of the same variable in the other equation in order to find the second variable

Variable a quantity whose value changes or varies. A variable could also be defined as a specific unknown in an equation. In algebra, letters often represent variables.

www.ingramcontent.com/pod-product-compliance
Ingram Content Group UK Ltd.
Pitfield, Milton Keynes, MK11 3LW, UK
UKHW050143280726
14058UKWH00006B/794

9 781524 928520